MOVING TOWARD SUSTAINABILITY:

Sustainable and Effective Practices for Creating Your Water Utility Roadmap

Table of Contents

Acknowledgments

Steering Group

Lisa Daniels
Pennsylvania Department of Environmental Protection Harrisburg, Pennsylvania

Todd Danielson
Avon Lake Regional Water, Avon Lake, Ohio

John Hollenbach
United Water of Pennsylvania and Delaware, Harrisburg, Pennsylvania

Andy Kricun
Camden County Municipal Utility Authority, Camden, New Jersey

George Martin
Greenwood Metropolitan District, Greenwood, South Carolina

Michael Mucha
Madison Metropolitan Sewer District, Madison, Wisconsin

Ron Poltak
New England Interstate Water Pollution Control Commission, Lowell, Massachusetts

Dan Roberts
City of Palm Bay Utilities, Palm Bay, Florida

Tom Sigmund
NEW Water, Green Bay, Wisconsin

Diane Taniguchi-Dennis
Clean Water Services, Hillsboro, Oregon

EPA Contributors

Jim Horne (lead)
Office of Wastewater Management

Jeffrey Fencil
Office of Ground Water and Drinking Water

Bonnie Gitlin
Office of Wastewater Management

Gary Hudiburgh
Office of Wastewater Management

Matt King
Office of Wastewater Management

Sonia Brubaker
Office of Ground Water and Drinking Water

John Whitler
Office of Groundwater and Drinking Water

This product was developed with assistance from **Rob Greenwood and Morgan Hoenig with Ross Strategic** (www.rossstrategic.com) **under Contract EP-C-11-009 with** the Office of Wastewater Management at the U.S. Environmental Protection Agency.

FOREWORD

Sustainable water and wastewater services are critical to providing the American public with clean and safe water and helping ensure the environmental, economic, and social sustainability of the communities these utilities serve. Many utilities across the country face tremendous challenges, such as aging infrastructure, climate changes, population growth, and competing resource priorities within the communities they serve.

As more and more utilities assume leadership roles related to community sustainability, resource recovery and conservation, sustainable economic development, and climate change, they must concurrently focus on long-term sustainability and bringing about meaningful change in their organizations and communities. For the past several years, the U.S. Environmental Protection Agency (EPA) has worked in collaboration with utilities, states, professional associations and others to help utilities across the water sector respond to these challenges.

In this spirit of collaboration, EPA is now issuing *Moving Toward Sustainability: Sustainable and Effective Practices for Creating Your Own Water Utility Roadmap*. Developed with extensive input from leading utilities, states, and professional associations, the purpose of this document is to assist utility leaders implement proven and effective practices over time to improve their operations and move toward sustainability, at a pace consistent with their needs and the needs of their communities. The practices are organized according to three separate business levels, using the well accepted Effective Utility Management framework supported by EPA and major professional associations. The document provides utility leaders with a cohesive structure to help them systematically address various challenges proactively and with confidence to create an individualized "roadmap" as it moves toward sustainable operations over time.

EPA believes the proven and progressive practices described in this document can help utilities:

- Save money by optimizing the planning and delivery of services to their customers;
- Ensure a reliable source of water consistent with customer needs;
- Use energy and water-efficient practices and technologies that foster water reuse, resource recovery, and green infrastructure;
- Become more resilient to short-term disasters and other longer-term climate related challenges; and
- Build greater understanding and support from decision-making bodies, customers, and other community stakeholders

Protecting our precious water resources and communities is a critical and ongoing challenge. Through this document and many other actions, EPA looks forward to helping utilities and communities address this challenge.

Nancy K. Stoner
Acting Assistant Administrator for Water

Introduction and Purpose of This Document

Sustainable water and wastewater services are critical to providing the American public with clean and safe water and helping ensure the environmental, economic, and social sustainability of the communities utilities serve. Utilities across the country face tremendous challenges, such as aging infrastructure, an aging workforce, increasing mandates, and competing priorities within the communities they serve.

The purpose of this document is to assist utility leaders with implementing proven and effective practices over time to improve their operations and move toward sustainability, at a pace consistent with their needs and the needs of their communities. It provides utility leaders with a cohesive structure to help them address various challenges proactively and with confidence. The practices described in this document reflect the lessons learned and the practical experience utilities have derived as they have improved their operations. A utility can use this document to identify specific opportunities for improvement and draw on the example practices to create an individualized "roadmap" to more sustainable operations.

This document is a continuation of a significant body of work led by the Office of Water at the U.S. Environmental Protection Agency (EPA) to promote actions to make water sector utilities of various sizes more sustainable and help ensure the sustainability of the communities they serve. EPA has collaborated closely with utilities, states, federal agencies, and other organizations in all of these efforts, including the following:

➤ Supporting effective utility management (EUM) based on a series of attributes of effectively managed utilities and keys to management success, as described in *Effectively Utility Management: A Primer for Water and Wastewater Utilities[1]*.

➤ Working with the U.S. Department of Agriculture (USDA) on a parallel initiative and associated publication targeted to rural and small systems – *The Rural and Small Systems Guidebook to Sustainable Utility Management[2]*.

➤ Finalizing a handbook to help utilities incorporate sustainability considerations into their existing planning processes – *Planning for Sustainability: A Handbook for Water and Wastewater Utilities[3]*.

---

[1] http://water.epa.gov/infrastructure/sustain/upload/2009_05_26_waterinfrastructures_tools_si_watereum_primerforeffectiveutilities.pdf

[2] http://water.epa.gov/infrastructure/sustain/upload/SUSTAINABLE-MANAGEMENT-OF-RURAL-AND-SMALL-SYSTEMS-GUIDE-FINAL-10-24-13.pdf

[3] http://water.epa.gov/infrastructure/sustain/upload/2009_05_26_waterinfrastructures_tools_si_watereum_primerforeffectiveutilities.pdf

➤ Developing and maintaining a tool to help water and wastewater systems develop asset management programs – the *Check Up Program for Small Systems (CUPSS)*[4].

➤ Developing guidance and tools to help water and wastewater systems better understand their energy usage and identify opportunities to increase energy efficiency – *Ensuring a Sustainable Future: An Energy Management Guidebook for Wastewater and Water Utilities*[5] and the *Energy Use Assessment Tool*[6].

The first two publications referenced above are particularly relevant to this document. Both publications present a set of utility management areas (e.g., Financial Viability) and keys to management success (e.g., Strategic Business Planning) and provide a cohesive, objective, step-by-step, self assessment framework for utilities to assess their strengths and areas for improvement. The practices provided in this document are presented to align with the management areas and keys to success from these publications, as they have been endorsed by EPA, water sector professional associations, and other federal agencies like USDA. Utilities that choose to use this document are encouraged to undertake a self assessment to gain a better understanding of which management areas and practices they wish to focus on first. Information on these assessment tools is available at http://water.epa.gov/infrastructure/sustain/watereum.cfm. Finally, the document reflects significant input from a group of leading utility and state managers. The Acknowledgements section of this document provides a list of these individuals. EPA is deeply appreciative of the contributions these individuals have made.

Industry professional associations and others have also provided significant leadership in this area. Examples include the *Energy Roadmap for Wastewater Utilities* developed by the Water Environment Federation (WEF); the *Utility of the Future Blueprint* developed by the National Association of Clean Water Agencies (NACWA), Water Environment Research Foundation (WERF), and WEF; and the *Sustainability Policy* of the American Water Works Association (AWWA).

Finally, this document complements, but does not duplicate these efforts. For example, an entire section of the document describes practices that are closely aligned with the directions set forth in the *Utility of the Future Blueprint*. Other practices in the document are consistent with the approach embodied in the WEF Energy Roadmap. Going forward, EPA will continue to work closely with industry and other partners to clarify how these various efforts complement each other and communicate this alignment to the water sector utility community at large.

---

[4] http://water.epa.gov/infrastructure/drinkingwater/pws/cupss/
[5] http://www.epa.gov/owm/waterinfrastructure/pdfs/guidebook_si_energymanagement.pdf
[6] http://water.epa.gov/infrastructure/sustain/energy_use.cfm

WHAT'S IN IT FOR ME?
WHY UTILITY LEADERS SHOULD CREATE A ROADMAP

Utility leaders, which include both managers and staff, are looking for practical, flexible, and user-friendly tools that can help them improve the day-to-day management of their operations. As more and more utilities engage in leadership roles on issues related to community sustainability, resource recovery and conservation, sustainable development, climate change, and environmental education, they must concurrently focus on long-term sustainability and bringing about meaningful change in their organizations and communities. This document provides a structure for creating a roadmap that can help utility leaders address these challenges and capture opportunities proactively and with confidence.

The proven and progressive practices described in this document can help utilities do the following:

➢ Save money by optimizing the planning and delivery of services to their customers.
➢ Better protect the environment by consistently meeting regulatory requirements.
➢ Ensure a reliable source of water consistent with customer needs.
➢ Recruit and retain a workforce necessary to ensure sustainable operations.
➢ Become more resilient to short-term disasters and other longer-term, climate-related challenges.
➢ Use energy- and water-efficient practices and technologies that foster water reuse, resource recovery, and green infrastructure.
➢ Build greater understanding and support from decision-making bodies, customers, and other community stakeholders.
➢ Work effectively with other community interests to implement innovative, watershed-based solutions and strengthen the local economy.

OTHER THINGS UTILITY LEADERS SHOULD KNOW ABOUT THIS DOCUMENT

- This document does not seek to define one roadmap for utilities to follow. It provides a flexible way for utilities to develop their own roadmap to meet the needs of their communities.
- Practices identified in this document are illustrative examples that reflect extensive input from leading utility and state managers from around the country, as well as EPA staff involved in various utility sustainability efforts.
- The practices, by design, are not comprehensive. They are progressive "practices with a purpose" to provide utility managers with practical examples that can help improve the sustainability of their operations over time.
- The practices can be scaled and implemented regardless of a utility's current capacity.
- This document organizes practices into three levels, which the next section describes in more detail. The levels can be viewed as a "progression model" that allows utilities to gauge where they stand in terms of adopting the practices identified, and also allows them to create a roadmap for improvement to meet their needs and the needs of their communities.
- These levels do not imply any judgment about a utility's current performance. This document acknowledges that utilities have different technical, financial, and managerial capabilities and local operating contexts.
- EPA encourages utilities to create a roadmap based on these practices over time and at a pace consistent with their current priorities, future goals, and the needs of their communities.

The Path to This Document

EPA hosted a two-day workshop in September 2012, which involved leading utility and state program managers to build on past work promoting sustainable utility management. The workshop's main objectives were to:

- Learn more about the challenges and opportunities facing water sector utilities as they strive for increased sustainability.
- Identify specific areas of effective utility practice that can anchor efforts to promote greater sustainability in the future.
- Identify gaps in current knowledge, tools, and practices along with collaborative efforts that can help fill these gaps.

During the meeting, the utility and state participants identified a preliminary list of practices that a wide range of utilities have used to make progress along a spectrum of sustainability. EPA developed this document using input from this meeting and various other sources. EPA worked with various associations and received extensive input from the steering committee members to produce this product.

Core Management Areas

Example practices presented in this document have been organized under the ten core management areas identified in the diagram below. These areas are based on the *Attributes of Effectively Managed Utilities* and *Keys to Management Success*, which EPA and major water sector associations support, and a similar framework that EPA and USDA developed for small systems. The core management areas presented horizontally (and shaded in blue) are based on the EUM Attributes and the EPA/USDA small systems framework. These core management areas are not presented in any particular order. Two of these core management areas – Product Quality and Operational Optimization and Customer Satisfaction and Stakeholder Understanding and Support - reflect a merger of EUM attributes for ease of presentation. The remaining two core management areas – Utility Business Planning and Performance Measurement and Continual Improvement – are drawn from the EUM Keys to Management Success. These core management areas flank the other areas to signal their importance to effective performance improvement. Good planning on the front end guides implementation of the practices and measurement and continual improvement on the back end enables adaptation and optimization as utility operating and external community and watershed priorities evolve over time.

Level 1 – Providing Adequate, Fundamental Services:

At this level, a utility is implementing practices that focus on meeting and maintaining compliance for all applicable regulations, ensuring adequate levels of operational resiliency, and implementing revenue and financing mechanisms that assure its mid- to long-term financial viability. The utility uses industry-accepted standard operating procedures (SOPs), proven and reliable technologies, and has clearly defined staff roles and responsibilities. The utility maintains a positive public image, cultivates an understanding of its operations and the value of its services with the community, is able to identify risks to high-consequence assets and plan for emergencies, and is capable of receiving and responding to customer and community concerns and complaints in a timely manner.

Level 2 – Optimizing Operations and Services:

In addition to providing basic services consistent with Level 1, a utility at this level is focusing on continual improvement and views optimizing its operations and services as central to mission success. The utility actively engages with its community to create operating conditions that are responsive to community needs and interests. The utility has established working relationships with neighboring systems as appropriate, has explicit performance improvement objectives and service levels, and actively seeks to ensure its operations support the community's economic and social well being. The utility has adopted sustainability as a core business principle and appropriately utilizes natural systems, like green infrastructure, in addition to other nonconventional technologies (e.g., decentralized approaches) and practices. The utility has started using processes for the internal recovery of energy, solids, and materials.

Level 3 – Transforming Operations and Services for the Future:

In addition to optimizing its current services, at this level, a utility is implementing practices consistent with many of the directions set forth in leading industry initiatives like the *Utility of the Future Blueprint.* The utility is employing practices that focus on managing treated wastewater and biosolids as valuable commodities, both to improve efficiency and as new revenue sources. The utility focuses on enhanced resiliency; acts as a leader in local watershed and community sustainability; and works actively with other local institutions to engage in community planning in order to help ensure economic, social, and environmental sustainability. The utility is a leader in and catalyst for economic development within the community; focuses on resource management and recovery; and works actively with others to promote full water cycle stewardship within its watershed to seek low-cost, high-return solutions. The utility also fosters an internal culture of innovation, collaborative development, and active engagement with its employees.

How to Read and Use this Document

The illustrative diagram below will help utilities understand how this document is structured in each of the ten core management areas. Utilities are also encouraged to take the following steps to help them most effectively use the practices presented in this document:

1. Assess your relative strengths and weaknesses in each of the core management areas. Prioritize those management areas you would like to initially focus on. EPA and other partners have developed tools to help utilities conduct such an assessment, which are available at
http://water.epa.gov/infrastructure/sustain/watereum.cfm.

2. Read through the practices in each of the three levels relevant to your selected management area(s) to determine the extent to which your utility has implemented specific practices (e.g., Fully Implemented, Partially Implemented, Not Implemented, Other).

3. Based on steps 1 and 2, identify specific actions you will take drawing on the example (or similar) practices, as appropriate. The Appendix to this document also includes a directory of resources designed to help you. These resources are organized around the Ten Core Management Areas presented in this document.

4. Develop an action plan that identifies the steps you will take to adopt the practice(s), who will be responsible, and a timeline for action.

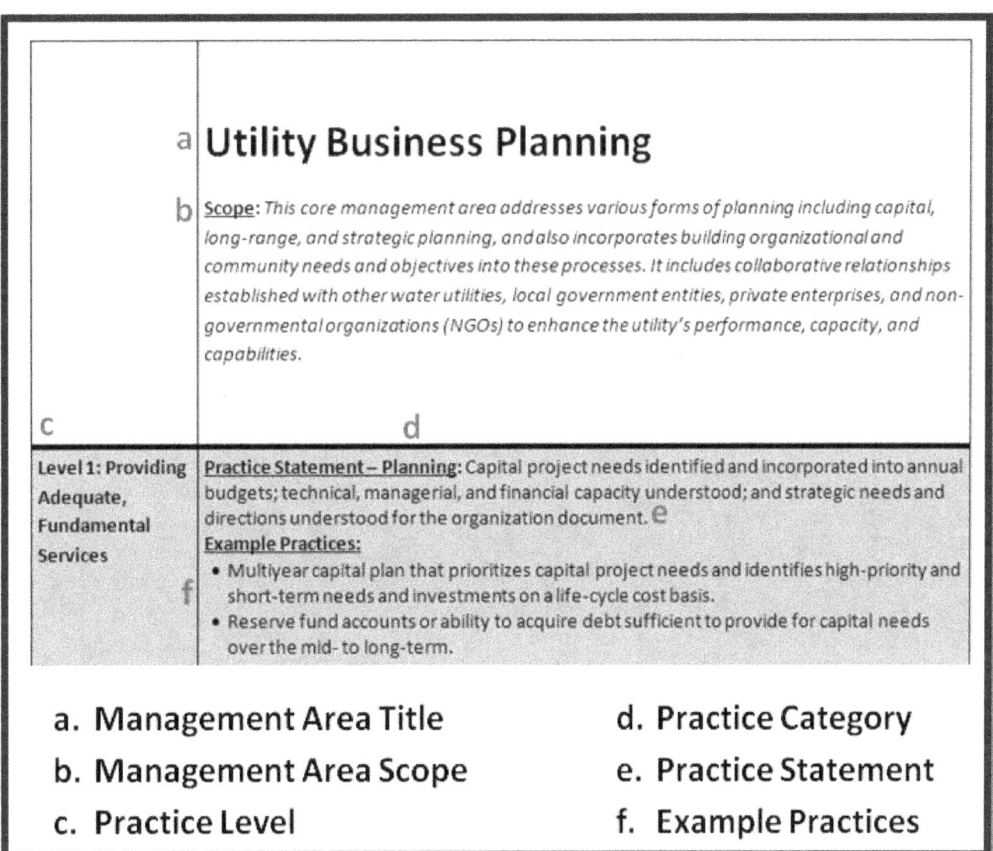

a. **Management Area Title**   d. **Practice Category**

b. **Management Area Scope**   e. **Practice Statement**

c. **Practice Level**   f. **Example Practices**

The Ten Core Management Areas and Example Practices

Utility Business Planning

Product Quality and Operational Optimization

Customer Satisfaction and Stakeholder Understanding & Support

Employee & Leadership Development

Financial Viability

Infrastructure Stability

Operational Resiliency

Water Resource Adequacy

Community Sustainability

Performance Measurement and Continual Improvement

Utility Business Planning

Scope: *This core management area addresses various forms of planning including capital, long-range, and strategic planning, and also incorporates building organizational and community needs and objectives into these processes. It includes collaborative relationships established with other water utilities, local government entities, private enterprises, and non-governmental organizations (NGOs) to enhance the utility's performance, capacity, and capabilities.*

Level 1: Providing Adequate, Fundamental Services	**Practice Statement – Planning:** Capital project needs identified and incorporated into annual budgets; technical, managerial, and financial capacity understood; and strategic needs and directions understood for the organization to document. **Example Practices:** • Multiyear capital plan that prioritizes capital project needs and identifies high-priority and short-term needs and investments on a life-cycle cost basis. • Reserve fund accounts or ability to acquire debt sufficient to provide for capital needs over the mid- to long-term. • Policies and procedures in place for capital project monitoring and reporting. • Capital planning and improvement program, which identifies how the capital plan fits into the utility's established policies, goals, and objectives and how the capital improvement process incorporates engineering and finance recommendations. • Technical performance, reliability, and maintainability guidelines and selection criteria to evaluate, compare, and identify priorities among capital project proposals. **Practice Statement – Partnerships/Collaboration:** Relationships established with water sector associations, regulators, and technical assistance providers to maintain awareness of and obtain support for implementing improvement opportunities. **Example Practices:** • Staff attendance at association conferences (state or national). • Relationships with regulators and other members of industry to solicit external input on compliance and performance. • Use of technical assistance services, such as those provided by states or water sector associations. • Municipal or state contracts used, as available (e.g., accessing state General Services bulk purchase contracts for such services as laboratory analysis).
Level 2: Optimizing Operations and Services	**Practice Statement – Planning:** Medium- and long-range (10 to 20 years) utility and community clean and safe water needs understood in a formalized, systematic, and transparent planning process. **Example Practices:** • Strategic and long-range planning that integrates utility priorities and goals with a basic understanding of community needs and interests. (Planning includes: goal setting reflective of utility and community priorities; explicit objectives and strategies in support of sustainability goals; alternatives analysis methods that integrate sustainability criteria; and financial strategy adequate to meet current and future needs). • Explicit business case processes, including cost-benefit analyses, for selecting all major capital investments.

	• Capital planning based on master plans that provide a vision for investments to correct existing deficiencies and meet future capacity needs. **Practice Statement – Partnerships/Collaboration:** Operational agreements in place with neighboring utilities to improve system performance, lower costs, or improve resilience. **Example Practices:** • Joint operating agreements for sharing administrative, maintenance, or other services (e.g., joint water and electricity meter reading). • Cross-training of staff from neighboring utilities. • Real-time control capabilities among neighboring systems that share collection or distribution networks. • Staff participation in association committees and leadership in research projects. • Proactive meetings with regulators and other key stakeholders to establish a pattern of positive interactions. • Relationships (formal or informal) with nearby facilities to enter into joint management agreements (e.g., sharing of certified operators, consultants, equipment, sample collection, lab analysis, water line replacement/repair, or bulk equipment purchases).
Level 3: Transforming Operations and Services for the Future	**Practice Statement – Planning:** Community priorities and broader watershed needs incorporated explicitly into utility decision-making; full internal alignment of utility vision, mission, and long-term planning exist; and sustainability is adopted as a core business principle. Utility vision, mission, investments, and operations integrated with other community departments. **Example Practices:** • Utility long-term strategy and mission, aligned with other community priorities, and integrated into long-range and capital planning. • Multi-attribute analysis to support incorporation of Triple Bottom Line principals and associated metrics explicitly into alternatives analysis for planning purposes. • Economic development incentive policy supported by cost-benefit, feasibility, risk, and uncertainty analyses. • Annual joint planning sessions with key community departments (e.g., Health, Transportation, Land Use and Planning, Parks, Economic Development, Arts), supported through efforts to encourage internal departmental adoption of planning results (e.g., change management training). • Community vision planning sessions (catalyze or participate in) to align department missions and develop a unified sustainability strategy for the community and watershed. • Scenario-based planning, or similar methods, to plan for and address complex uncertainties such as impacts from a changing climate. **Practice Statement – Partnerships/Collaboration:** Proactive relationships established with external parties critical to the organization's operating environment and core mission. Strategic relationships for short- and long-term collaboration and in support of alternative services also established. **Example Practices:** • Lead or participate in coordination among local drinking water, wastewater, and stormwater utilities to integrate management strategies and long-term planning. • Define policy criteria for making various contributions to, or investments in, partnership arrangements. • Fund and operate energy generation technologies and resource recovery with public-private partnerships. • Share information and coordinate with land use agencies on watershed planning.

Product Quality and Operational Optimization

Scope: *This core management area includes compliance with regulatory requirements, energy and materials usage (chemicals and residuals), technology, and standard operating procedures (SOPs).*

Level 1: Providing Adequate, Fundamental Services	**Practice Statement – Compliance/Performance Standards:** Compliance obligations understood and consistently met, and applicable industry performance standards understood and adopted. Compliant industrial pretreatment program in place, and residuals compliant with applicable regulations. **Example Practices:** • Water quality sampling and process control monitoring SOPs to support proactive recognition of possible compliance problems for all Safe Drinking Water Act (SDWA) and Clean Water Act standards (e.g., biochemical oxygen demand and total suspended solids monitoring in wastewater treatment plant discharge mixing zones; and drinking water chlorine residuals monitoring). • Basic compliance metrics established and tracked for wastewater treatment effectiveness rate and relevant SDWA requirements. • Program for identifying and incorporating new regulations, drinking water standards, or discharge limits. • Industrial pretreatment program certified by state environmental protection agency in place. • Applicable operator certification requirements program in place. • Record-keeping and reporting requirement SOPs. **Practice Statement – Operational Improvement and Maintenance*:** Opportunities to improve operations undertaken as they are identified and time and resources allow. (The utility reacts to information provided from outside sources relating to improvement opportunities). ** Refer to the Infrastructure Stability section for maintenance details.* **Example Practices:** • Odor control measurement, monitoring, and response program. • Leak detection and repair program. • Optimization targets for use of labor, chemicals, and residuals. • Valve exercise and maintenance program. • Cross-connection control program. • Water meter repair/replacement program. **Practice Statement – Energy/Materials:** Basic energy efficiency opportunities identified and implemented. Basic chemical safety procedures in place. **Example Practices:** • Energy use assessment. • Chemical Right to Know program for all employees. • No- or low-cost energy efficiency practices adopted (e.g., variable frequency drives wherever possible, low-energy lighting, etc.). **Practice Statement – Technology:** Use fully characterized and proven technologies and management systems. All plant process control systems functioning effectively and leading to full compliance for drinking water, wastewater discharges, and solids/residuals.

	Example Practices: • Adequate sizing and maintenance for wastewater treatment facilities to ensure consistent permit compliance. • Outside resource (e.g., regulatory agencies or consultants) usage to supplement in-house capacity and assist with decision making and implementation. • Reliable disinfection methods operating consistently with permit compliance.
Level 2: **Optimizing** **Operations and** **Services**	**Practice Statement – Compliance/Performance Standards:** Beyond compliance service levels and practices adopted. "Outstanding performance" achieved in sanitary surveys. Proactively contribute to beyond compliance voluntary standards development. **Example Practices:** • Voluntary optimization standard adoption (e.g., Partnership for Safe Water, EPA's Area Wide Optimization Program). • Measurement, monitoring, and corrective action procedures of all key process units for detecting and responding to compliance "near misses." • Written SOPs for critical utility operational functions. • Participation in water research activities, such as "pilot projects" or other testing for new, voluntary standards. • Water quality monitoring (source and distribution) beyond regulatory requirements. • Industrial user recognition program for beyond compliance performance (e.g., reduction of caustic chemical discharges to sewer system). **Practice Statement – Operational Improvement and Maintenance*:** Create and implement an operational optimization plan and program. Utility seeks to improve performance beyond compliance requirements. **Refer to the Infrastructure Stability section for maintenance details.* **Example Practices:** • Standardized equipment and functions (e.g., uniformly outfitting trucks). • Resource optimization targets (e.g., created and monitored for operation and maintenance (O&M) costs per population served, cost of customer billing per service connection, water collected or treated per employee). • Distribution system pressure monitoring. • Real-time on-line monitoring for key parameters (e.g., pH, chlorine residual, etc.) to ensure optimal operation. • Available process equipment capability optimized (e.g., for effluent quality to the maximum extent practicable). • Distribution system water age management program. • Optimization programs to improve treatment efficacy and water quality, and to reduce chemical usage, energy costs, and sludge volume. **Practice Statement – Energy/Materials:** Energy management plan in place; proactive industrial pretreatment program, including pollutant trackback and pollution prevention program in place; opportunities for reductions of chemical usage identified and implemented; plan optimized to maximize residual capture and residuals used for beneficial reuse to the maximum extent possible; and utilization of plant effluent for process water needs. **Example Practices:** • System-wide water and energy audits conducted. • Explicit energy optimization actions and goals (e.g., energy reduction targets adopted and tracked). • Trackback program for pollutants of concern.

	<ul><li>Chemical monitoring systems to minimize probability of unnecessary overdosing.</li><li>Optimized sludge thickening and dewatering equipment to maximize solids capture capacity.</li><li>Vigorous manufacturer and other sources of inorganic pollution outreach and education to help utilities implement best pollution prevention practices.</li><li>Water optimization program (including re-use strategy and targets, water loss control program, and customer water audit program).</li><li>Take-back program promotion (e.g., for unused pharmaceuticals).</li></ul> **Practice Statement – Technology**: Innovative technology adoption capacity in place. Practices to help reduce the sewage and stormwater burden on sewer lines and sewage treatment plants, and reduce the need for capital upgrades in place. **Example Practices:**<ul><li>Green infrastructure for stormwater management source control (e.g., rain gardens, permeable pavement, and green parking lots).</li><li>Advanced treatment technology evaluation in equipment replacement and capital improvements.</li><li>Secondary treatment and filtration automated process systems (e.g., membrane treatment, UV treatment).</li><li>Infiltration/inflow removal and impermeable surface reduction.</li><li>Liquefied Petroleum Gas (LPG)-fueled vehicles.</li><li>Practices to help mitigate groups of contaminants or contaminant precursors from drinking water.</li></ul>
Level 3: Transforming Operations and Services for the Future	**Practice Statement – Compliance/Performance Standards:** Compliance assurance capacity expanded to address new regulatory areas that come with marketing and sale of resource recovery products or utility proprietary services and tools. Proactive engagement with regulators, other utilities, and watershed participants to improve compliance performance and establish more effective performance expectations. **Example Practices:**<ul><li>Networking with energy and consumer product regulators to understand regulatory requirements and establish proactive working relationships.</li><li>Proactive input to regulatory agencies in establishing new compliance levels.</li><li>Assistance to other utilities to improve their performance (e.g., equipment loans, knowledge, and other resources).</li><li>Active and effective environmental advocacy beyond the bounds of the utility's facilities.</li></ul> **Practice Statement – Operational Improvement and Maintenance***: Advanced optimization methods and practices deployed. Integrate data system for automated real-time control and optimize utility and other community systems. **Refer to the Infrastructure Stability section for maintenance details.* **Example Practices:**<ul><li>Real-time system control. (For example, traffic of peak flows or peak demands is controlled to manage the amount of water in the system by shuttling flow between treatment plants).</li><li>Standard data infrastructure across interdependent utility operations.</li><li>Utility Geographic Information System (GIS) layer integration across community service departments.</li><li>Data and automated systems integration in the context of shared water utility operations (optimize existing infrastructure).</li><li>Real-time monitoring of distribution system water quality.</li><li>Dashboard system to track key indicators of importance to the utility (e.g., customer service, treatment quality, cost).</li></ul>

- GIS, hydraulic model, Supervisory Controls and Data Acquisition (SCADA), and customer information system full integration.

Practice Statement – Energy/Materials: Internal energy resource recovery adopted; deployment of enhanced energy generation approaches to meet 100 percent of energy needs (i.e., "net zero"); and initial advanced materials recovery.

Example Practices:

- Alternative and renewable energy sources for plant operations (e.g., implementing solar power, wind power, or hydroelectric power; biogas for space heating; and conversion of biogas to electricity).
- High-strength waste acquisition and digestion (e.g., Fats, oils, and grease (FOG) or organics recovery from street sweeping).
- Partnerships for acquisition and marketing of energy (e.g., electric and gas utility partnerships, high-strength waster partnerships).
- Resource recovery for nitrogen, phosphorus, organic material, and possibly precious metals.
- Production and supply of different water qualities (e.g., local stormwater used for toilet flushing).
- Energy recovery in treatment and distribution systems (e.g., sewer geothermal [using heat exchange technology with sewer wastewater], smart meters, and hydrokinetic turbine applications).

Practice Statement – Technology: Advanced, distributed technology deployment. Capability to explore (pilot test) and deploy emerging technologies.

Example Practices:

- Collection system used as an extension of treatment.
- Pilot projects (internally led) for testing promising technologies.
- Robust connection to research and development affiliates (e.g., foundations, university partnerships).
- Microfiltration equipment.
- Natural treatment systems to facilitate biological removal (e.g., enzymes or catalysts).
- Decentralized supply and treatment evaluated and performed on a watershed, water-quality basis.

Customer Satisfaction and Stakeholder Understanding & Support

Scope: *This core management area includes engagement and education efforts, customer feedback and response mechanisms, promotion and public relations, and participation in public events.*

Level 1: Providing Adequate, Fundamental Services	**Practice Statement – Community Engagement and Participation:** Utility is a visible member of the community and plays an active role in community events. **Example Practices:** • Participation in community events (e.g., staff organizing to volunteer at fundraising events). • Community organizations or sports team sponsorships. • Community event booth sponsorships. • Volunteer for community events (e.g., organize watershed cleanups). • Public information event sponsorships (e.g., Water Week). • Utility open house events. **Practice Statement – Public Relations, Education, and Promotion:** Basic information about the utility is readily accessible and understandable to community members. Emergency response communications proactively prepared. **Example Practices:** • Website includes information about major undertakings, and important documents are available. • Periodic explanatory customer flyers. • Clear, visible signs for construction activities. • Multiple-language utility documents consistent with community profile. • Emergency event public communications templates for media contacts (e.g., create standardized text for boil water notices). • Annual consumer confidence/water quality report. **Practice Statement – Customer and Stakeholder Feedback and Response:** Customer complaint and response mechanisms are in place. **Example Practices:** • Customer complaint response time targets. • Key community stakeholder opinion leaders list and schedule for outreach (e.g., phone call, informal meeting). • Customer information system to store billing information, service requests, and all resolutions. • Customer complaint receipt and response capability.

Level 2: Optimizing Operations and Services	**Practice Statement – Community Engagement and Participation:** Community organizations and members engaged as full partners in utility plans and operations.
	Example Practices:
	• Two-directional engagement with community members (e.g., through social media or advisory councils).
	• Media approaches geared to needs of different generations.
	• Community engagement forums to understand critical values, set utility goals, and review infrastructure alternatives.
	• Public participation in the planning, budget, and performance management results processes.
	Practice Statement – Public Relations, Education, and Promotion: Focus efforts to increase community understanding of the utility, the benefits from its functions and services, and the requirements for operating sustainably. Utility is viewed as a leader and critical, trusted player in the community and citizens have a strong working knowledge and acceptance of the requirements for operating sustainably.
	Example Practices:
	• Value of water and water services educational brochures and public education campaign.
	• School outreach programs (e.g., K–12 classroom presentations or local school science program water-related curriculum).
	• Electronic budget documents and comprehensive annual financial report (CAFR) on the utility's website, including a concise summary and guide to the key issues of the operating and capital components.
	• Executive director role focused on external communication and relationship building.
	• Annual utility performance report based on service level commitments and using common language and illustrative examples (e.g., "we saved 10,000 barrels of oil this year through our energy conservation efforts").
	• Annual plain language report on "Capital Facilities" for elected officials and the general public that describes the condition and plans for asset replacement and renewal.
	Practice Statement – Customer and Stakeholder Feedback and Response: Feedback actively solicited and mechanisms for understanding and improving satisfaction and support are in place. Utility has established trust relationships with key community opinion leaders and stakeholders and maintains regular interactions to provide updates and stay abreast of external needs and interests.
	Example Practices:
	• Customer satisfaction surveys.
	• Customer feedback focus groups.
	• Customer complaint management system to monitor and respond to complaints.
	• Methods for incorporating customer feedback into change management processes.
	• Regular regulator contact to establish a positive, proactive relationship (e.g., creating an understanding for the best allocation of funds).
	• Focused outreach and information sharing with the financial community, particularly in the context of bond ratings.
Level 3: Transforming Operations and Services for the Future	**Practice Statement – Community Engagement and Participation:** Utility is a catalyst to create networks among community departments, organizations, and stakeholders in support of watershed and community-wide sustainability improvements. The utility exerts leadership among community and regional stakeholders regarding watershed and water quality improvements.
	Example Practices:
	• Watershed forum sponsorship for developing community-wide water sustainability and economic development strategy.

- Community event co-sponsorship with other community organizations related to water or watershed protection to increase the public visibility of water.
- Expanded utility public engagement scope to include full range of watershed participants.
- Coalitions with NGOs to leverage resources in support of utility, watershed, and community sustainability initiatives.

Practice Statement – Public Relations, Education, and Promotion: Utility has positioned itself as a leader in community sustainability, and through this leadership it influences other community organizations to follow suit. Utility has an outreach strategy designed to support its efforts to market new services and products effectively.
Example Practices:
- Stakeholder and customer surveys to test knowledge of utility-related issues and outreach and education efforts targeted to fill knowledge gaps.
- Risk management communication to the public in support of innovation (increase public tolerance for service failure or increased costs).
- Branded utility services and products.
- Utility culture includes clear articulation and communication as a part of the organization's brand.

Practice Statement – Customer and Stakeholder Feedback and Response: Acceptance of utility by customers and stakeholders as a valuable community resource that can and should be involved in the provision of services and products beyond clean and safe water in areas like economic development.
Example Practices:
- Services marketing to prospective customers about how the utility could better serve them (e.g., businesses considering moving to the area, local producers of high-strength waste).
- New stakeholder engagement around nontraditional services (e.g., energy production).
- Economic corridor identification and focused planning and design for water and wastewater infrastructure support.
- Utility GIS asset leveraging to accomplish enterprise support for business development (i.e., map visualization/layering of building zones, future land use, "smartzones," enterprise zones, "HUBzones," census tracts and blocks, present and future transportation networks, etc.).
- Utility marketing and graphic design asset leveraging to support municipal government's development of a community guide promoting the benefits to business of locating in a sustainable community that EUM supports.

Employee & Leadership Development

Scope: *This core management area includes organizational structure, workplace culture, institutional knowledge, succession planning, and employee development opportunities.*

Level 1: Providing Adequate, Fundamental Services	**Practice Statement – Workplace Culture:** Clarity is established for all job responsibilities and functions. Sufficient workforce, with necessary training, is in place in all staff functions. **Example Practices:** • Written job requirements and descriptions for all staff functions. • Organizational roles chart available and up-to-date. • Annual performance reviews with written feedback. • Periodic employee celebrations of organizational performance success (team accomplishments). • Formal ethics policy. **Practice Statement – Recruitment, Retention, and Succession:** Needed skills and expertise are documented, used to screen applicants, and effectively communicated to new hires; reasons for employee turnover are understood. **Example Practices:** • Interview process with standardized questions tailored to the position. • Employee turnover statistics tracked and evaluated. • Formalized and standardized new hire orientation template listing materials to provide, key topics to review, and skills to impart. • Critical position identification and characterization for recruitment purposes. • Equal opportunity hiring policy in place to facilitate workforce diversity. **Practice Statement – Development Opportunities:** Program in place to support and enable staff to acquire and maintain required professional certifications. **Example Practices:** • Certification needs and opportunities identification and inclusion in annual budget. • Operator training and education reimbursement, and leave allowance for needed certifications. • Merit pay increases for acquired certifications. • Staff cross-training across functions and departments to augment system resiliency.
Level 2: Optimizing Operations and Services	**Practice Statement – Workplace Culture:** Performance expectations are explicitly established and tied to compensation. Employees are encouraged to provide ideas and feedback to improve operational and administrative performance. Desired organizational culture is clearly defined and communicated to employees. **Example Practices:** • Written job descriptions with explicit, systematic performance evaluation metrics and standards. • Skill level and expertise requirement articulation with link to pay stratification/raises. • Employee suggestions for improvement program with management responsiveness. • Employee awards and recognition program linked to creative thinking and continual improvement efforts related to achieving organizational goals.

- Employees encouraged to make decisions and take independent actions that fall within organizational guidelines.
- Employees engaged in annual organizational goal and long-term strategic planning processes.
- Executive management informal workforce engagement (e.g., treatment plant walk-through and regular opportunities to meet with staff at all levels).
- Employee development plans based on performance and skill evaluations, as well as employee professional goals.

Practice Statement – Recruitment, Retention, and Succession: Current and future workforce requirements are understood, with proactive efforts made to attract and retain highly qualified staff.
Example Practices:
- Key skill set identification for strategic positions and specific media and other venues used to implement recruitment strategy (e.g., state-managed matching of certified professionals to utility jobs used to solicit employees to acquire those skills).
- Exit interviews to understand reasons for separation and to identify opportunities for improving employee motivation and loyalty.
- Formal retention management plan used to help identify the most critical employees to retain.
- Full on-boarding program for new hires: resources, communications, organizational culture, training, welcoming activities, and guides (mentor).
- Workforce demographics documentation and retirement projections (incorporated into succession and recruitment plans).
- Current workforce skills inventory, future workforce skills projections, and needed skills training.

Practice Statement – Development Opportunities: Explicit professional development program for staff and management is in place, including incentives for personal improvement and activities designed to increase the "bench depth" of staffing. A structured training program establishes ongoing requirements and opportunities for professional development, and staff is encouraged to engage in broader water sector professional development opportunities.
Example Practices:
- Tuition reimbursement program to incentivize professional development consistent with organizational needs and goals.
- Staff rotation to other utilities or functions within the utility for cross-training and mentoring.
- Critical workforce competencies analyses by management (with subsequent targeted training programs emphasizing use of continuous improvement tools).
- Broad-based leadership and management skills training conducted annually for formal and informal leadership positions and opportunities.
- Additional leave time (especially in leadership roles) to promote membership in professional organizations.
- Explicit training program for all staff covering core organizational functions: managerial and supervisory, professional/technical, business practices, safety, compliance, IT systems, customer service, interpersonal skills, and executive development.
- Formal leadership training focusing on: vision, mission, values; organizational culture; human resource policies and bargaining unit agreements; interpersonal skills; conflict resolution; problem solving and decision making; budgeting and budget management; performance appraisal; leadership; diversity; and grievance.

Level 3: Transforming Operations and Services for the Future	**Practice Statement – <u>Workplace Culture</u>:** Performance management system is in place that explicitly aligns employee incentives, compensation, and performance expectations with the organization's mission, objectives, and business plan. A strong participatory culture exists with staff members, who are encouraged to share ideas and take measured risks. A culture of innovation, collaborative development, and active employee engagement is established and actively enhanced by the entire workforce. **<u>Example Practices:</u>** • Standing collaborative forums with collective bargaining units, as appropriate (all employees included). • Individual employee annual performance plans with direct links to business plan objectives. (The plan review is conducted several times each year between the employee and supervisor). • Goal-sharing bonus programs for employees that reward employees for meeting their performance goals, which also help the organization meet its goals. • Organizational sustainability principles, commitments, and expectations incorporated into day-to-day operations. • Budget support for "innovation proposals" (removing barriers to creative thinking, and developing systems for failure tolerance). **Practice Statement – <u>Recruitment, Retention, and Succession</u>:** An active commitment to attracting and developing new employees exists, as well as the capacity to understand and track employee satisfaction and engagement. Critical skill and expertise requirements are understood and plans are in place to ensure their timely replacement. Skills and expertise requirements are actively updated to keep pace with operational innovations and business strategy requirements. Employee motivation and retention systems use multiple avenues to achieve objectives. **<u>Example Practices:</u>** • Education and recruiting partnerships (e.g., internship programs) through ongoing relationships with high schools, community colleges, and universities. • Mission and vision incorporated into branding efforts to support recruiting and retaining high-quality talent. • "Pay for performance" systems to provide incentives for high-performing staff. (The process used is well understood by staff, and is tied to goal attainment and specific performance criteria). • Annual employee satisfaction surveys to identify gaps and opportunities for training and employee development. • Job satisfaction focus groups with management to address issues identified in annual employee survey. (Solutions are determined through a collaborative approach between management and the workforce). **Practice Statement – <u>Development Opportunities</u>:** Establish a plan for cultivating the expansion of staff skills and expertise consistent with altered operating and technology environments relative to "utility of the future" operational demands. **<u>Example Practices:</u>** • Friendly utility-to-utility competition with neighboring systems in support of voluntary self-improvement programs. • Integrated workforce development curriculums. • Alternative management skills training (e.g., collaborative partnership development). • Emerging opportunities for skill-building collaboration between staff and management. (Employees work in conjunction with supervisors to develop and implement plans to build skills needed to support emerging opportunities that the organization faces). • New employee skill set sharing (e.g., identifying new skill sets and sharing them through coaching).

Financial Viability

Scope: *This core management area includes rates that reflect the full cost of service, accounting practices, fees, reserves, debt management, and the creation of additional revenue streams.*

Level 1: Providing Adequate, Fundamental Services	**Practice Statement – Accounting, Auditing, and Financial Reporting:** Controls and timely financial statements (issued as part of a CAFR) are in place reflecting Generally Accepted Accounting Principles with internal and independent audits conducted to ensure the system's integrity. **Example Practices:** • Policies on required level of working capital. • Targets for days of operating expense coverage. • Accounting policies and procedures, formally documented and consistently applied, to provide for the reporting of fraud or abuse and questionable accounting or auditing practices. • Policies for internal control procedures over financial management (periodically evaluated with auditors). • Policies and procedures on how to account for disaster-related reimbursable costs, and methods to track emergency incident expenses to facilitate cost reimbursement activities. • Policies and procedures on managing capital assets and on capitalization thresholds. **Practice Statement – Budget and Fiscal Policy:** Operational and capital funding needs understood and translated into rate and fee requirements, with a strategy and policies in place to maintain rates and fees at necessary levels. **Example Practices:** • Rate studies (to link rates to system needs). • Built-in, gradual, annual rate increases. • Operating reserve fund. • Financial policy development (includes financial planning policies, revenue policies, and expenditure policies), adoption, annual review, and communication to the governing board. • Explicit budget process and forecasts (used when preparing the utility's budget). • Procedure for program expenditures that exceed an established limit from the approved budget. **Practice Statement – Debt Management:** Annual budget accommodates financing capital reserves at levels needed to support capital replacement. Debt payments are made on a timely and cost-effective basis. **Example Practices:** • Capital reserve fund. • Comprehensive written debt management policy that addresses debt limits, debt structuring practices, debt issuance practices, debt management practices, and using derivatives. • Policy and procedures to ensure fiduciary responsibilities. • Policy and procedures for investing bond proceeds to ensure that legal and regulatory requirements are met, fair market value bids are received, and issuer objectives for various uses of proceeds are attained.

	• Capital plan financial feasibility analysis to identify financing methods and funding sources, and to assess funding availability and constraints. **Practice Statement – Procurement and Inventory:** Clarity and controls are established for maintaining efficient and consistent purchasing and inventory management. **Example Practices:** • Purchasing policy that standardizes procedures for ordering, accepting, or rejecting materials and services. • Operating inventory of supplies (defined, maintained, and updated to meet the needs of the utility operations). • Policy that determines when the procurement of goods and services requires a formal contract. • Emergency procurement policy to allow the expenditure of funds to support response and recovery activities after an emergency.
Level 2: Optimizing Operations and Services	**Practice Statement – Accounting, Auditing, and Financial Reporting:** Fiscal performance expectations are created with policies, practices, and targets in place to drive performance, create accountability, and support transparency. **Example Practices:** • Formal audit committee to provide independent review and oversight of the financial reporting process, internal controls, and connect with independent auditors. • Policy for level of unrestricted fund balance that should be maintained. (The target is analyzed and set based on particular characteristics and criteria of the utility and includes: transfers, cash cycles, customer profile, control over revenue, asset age and condition, volatility of expenses, control over expenses, and debt position). • Policy for target level of working capital. (The target is analyzed and set based on particular characteristics and criteria of the utility and includes: transfers, cash cycles, customer profile, control over revenue, asset age and condition, volatility of expenses, control over expenses, and debt position). • Financial statements with management's department-level discussion and analysis. • Mechanism to permit the confidential, anonymous reporting of concerns about fraud or abuse and questionable accounting or audit practices to the appropriate responsible parties. • Budget to actual comparisons in the audited basic financial statement. • Formal internal audit function. **Practice Statement – Budget and Fiscal Policy:** Cost of service is understood with rates and fees established accordingly, while revenue needs over the mid- to long-term are understood and rate impacts to customers are explicitly managed. **Example Practices:** • Payment assistance programs for disadvantaged households. • Cost of service studies. • Rate model to support current and future rate needs determinations. • Affordability criteria (and tracking the impact of bills on customers), with appropriate considerations for disadvantaged households. • Regularly monitor and periodically update major revenue and expenditure that extends at least three to five years beyond the budget period. • Separate rates for internal and external customers for designated goods or services according to financial objectives, equity, efficiency, and administrative feasibility.

	• Long-term financial plan, which looks at least five to ten years into the future; considers all appropriated funds; and is updated based upon debt position and affordability analysis, with strategies to achieve and maintain financial balance with a scorecard of key indicators of financial health that is visible to the public. **Practice Statement – Debt Management:** Policies and procedures are in place to ensure effective debt management, maintenance of a competitive bond rating, and capital needs are understood and addressed for the mid- to long-term. **Example Practices:** • Debt-to-equity targets for capital spending. • Strategic financial plans to avoid rate spikes. • Strong master bond resolutions, such as covenants, which prescribe coverage ratios. • Proactive bond refund evaluations (by bond counsel and financial advisors) to achieve interest cost savings; remove or change burdensome bond covenants; or restructure the stream of debt service payments to avoid default, or an unacceptable tax or rate increase. • Level of disclosure to bond holders' analysis (addressing the utility's pension funding obligations) with input from legal counsel and financial advisors. • Bond proceed investment risk analysis (to identify actions to mitigate risks). **Practice Statement – Procurement and Inventory:** Ability to track specific utility property location and usage is in place and standardized processes for disposition of property created. **Example Practices:** • Property disposal procedures. • Procurement of property and equipment records (including land, buildings, expendable items, installed property, uninstalled property, equipment, vehicles, and personal wear items owned by or assigned to the agency above a specified value). • Perpetual inventory system.
Level 3: Transforming Operations and Services for the Future	**Practice Statements – Accounting, Auditing, and Financial Reporting:** Fund balances are supported by targets, regular monitoring, and actions taken to maintain expected balances over time. Full transparency of financial performance and accounting practices is provided. **Example Practices:** • Fund balance replenishment rate targets. • Aggregated or consolidated presentations to supplement the CAFR (customized for a broad general audience to understand the utility's financial position in an objective manner). • Web-accessible financial statements. • Systematic effort to annually track and manage controlled capital assets at the department level. • Physical inventory of tangible capital assets (periodically performed and all assets are accounted for, at least on a test basis, no less than once every five years). **Practice Statement – Budget and Fiscal Policy:** Rates and fees are viewed as more than a means to fund operations, with focus emerging on using rate and fee coverage and structures to influence customer and community behavior in line with utility sustainability objectives. Cost-sharing strategies for a range of service provisions have been explored, and those making operational and financial sense adopted. Revenue strategy incorporates an effort to diversify utility revenue sources beyond those associated with conventional treatment services. **Example Practices:** • Conservation rate structures. • Process for designing other post-employment benefits to ensure sustainable funding approach is in place.

- Full cost of providing service estimates (calculated and considered in the basis for setting charges and fees—full cost incorporates direct and indirect costs including design, O&M, overhead, replacement, and charges for using capital facilities).
- Personnel tracking system (to accurately project budget and payroll based on the estimate of budgeted positions for the year that includes consideration of vacancy adjustments, collective bargaining, inflation, and compensation).
- Internal service rates (established for operations such as information technology, payroll, motor pool budgeting, legal, accounting, and human resources).
- Intellectual property development and marketing (e.g., watershed analysis models).

Practice Statement –Debt Management: Capital investment and debt management strategy adopt an explicit risk management posture focusing on managing investments to preserve and create new options in the future. Capital funds are created and managed to provide resources for technology innovation and partnerships are established to attract capital and risk share.
Example Practices:
- Avoiding over-investment in capital strategy. (Remain nimble by providing flexibility to take advantage of new technologies as they emerge, as well as managing for uncertainty).
- Investor relations program (to provide full and comprehensive disclosures of annual financial, operating, and other significant information in a timely manner consistent with federal, state, and local laws).
- Innovative technology deployment funds (specifically set aside for this purpose).
- Infrastructure and technology cost- and risk-sharing mechanisms (e.g., public-private partnerships for biogas development).
- Investor relations information dissemination (provided on website to the municipal securities market regarding utility debt, financial condition, and other related information).

Practice Statement: Procurement and Inventory: Capability is established to support disclosure requirements associated with asset value and depreciation over time. Procurement activities are integrated with the utility's sustainability commitments to ensure purchasing is aligned with utility sustainability performance expectations.
Example Practices:
- Financial reporting procedure for capital and infrastructure assets consistent with Governmental Accounting Standards Board Statement Number 34, as applicable.
- Property management system (well structured for managing property owned or used by the agency that provides for identifying, labeling, and recording existing capital assets, and is updated as assets are added, transferred, replaced, or destroyed).

Infrastructure Stability

Scope: *This core management area relates to the management of infrastructure and other physical assets.*

Level 1: Providing Adequate, Fundamental Services	**Practice Statement – Infrastructure O&M:** Maintenance is undertaken as performance deficiencies dictate, backed up by an explicit maintenance management system for assets above and below ground. **Example Practices:** • Work orders linked to asset inventory. • Ongoing training and certification/licensing requirements for maintenance staff. • Record retention of asset maintenance performed (e.g., work order system in place to keep maintenance records). • Estimated useful life and depreciation policy (in accordance with generally accepted accounting principles). • Manufacturer's recommended maintenance regimens followed for all equipment. • Emergency maintenance procedures. **Practice Statement – Asset Management:** Asset management program basics are understood and recognized as important for EUM. Assets are inventoried with information stored in a standalone database. **Example Practices:** • Critical infrastructure and assets inventory (includes original cost, with new assets recorded at the time of purchase and retired assets removed from inventory). • Critical infrastructure asset mapping (e.g., GIS-located mains, hydrants, valves, services, and tanks). • Photographic documentation of assets to compare baseline conditions to pictures taken after the asset is impacted during an emergency event.
Level 2: Optimizing Operations and Services	**Practice Statement – Infrastructure O&M:** Proactive, risk-based maintenance, repair, and replacement are used and technology and equipment standardization efforts undertaken with specific actions under way to improve the efficiency of infrastructure repair and rehabilitation. **Example Practices:** • Level of service and planned maintenance targets (with performance measures in place and tracked as part of the budget process). • Collection system line inspection and cleaning (for sanitary sewer overflow prevention). • Joint maintenance partnerships with other systems. • Root cause analysis for failures used to drive maintenance of asset decisions. • Underground asset replacement or restoration innovative solutions or restoration (i.e., water main relining, ice picking, bursting). • Critical spare parts inventory and all equipment either in operation or in fully ready standby mode. • Visible SCADA and GIS to enhance O&M. • Underground replacement/repair coordination with other projects (e.g., street paving).

	Practice Statement – Asset Management: Asset management is adopted as a core utility business function, guided by explicit service levels. Complete asset inventory and asset condition assessment is combined with the capability to make infrastructure repair and replacement decisions on a managed risk basis. **Example Practices:** • Regular asset performance assessments. • Asset full life-cycle cost estimates and depreciation studies to determine expected life cycles. Program to replace underground infrastructure on either a regular cycle (e.g., 100-year life cycle) or at the asset depreciation rate. • Hydraulic modeling analysis for the design of new and replacement infrastructure. • Condition assessment, monitoring, and failure analysis of infrastructure assets. • Service interruption tracking conducted relative to established levels of service targets. • Short-term and long-term asset management and capital plan supported by commitments for necessary funding. • Condition/functional performance standards defined for each type of capital asset.
Level 3: Diversifying Operations and Services for the Future	**Practice Statement – Infrastructure O&M:** Commitment to utility and community sustainability is explicitly incorporated into infrastructure investment and management efforts. Collaborative partnerships are sought and used to improve operational efficiency, manage risk, and improve resiliency. **Example Practices:** • Infrastructure project rating systems (e.g., Institute for Sustainable Infrastructure). • Alliance partnerships for infrastructure development. (Avoid low-bid constraints). • International Organization for Standardization (ISO) certification for asset management. • Innovative solutions to leverage capital markets for infrastructure sustainability (e.g., equipment manufacturer partnerships, public-private contractual arrangements, design build operate). • SCADA integrated with Computerized Maintenance Management System and GIS Enterprise system for optimizing asset management (e.g., tie maintenance and repairs to system assets). **Practice Statement – Asset Management:** Capital and natural resource asset diversification are used to manage risks and boost resiliency, while collaborative partnerships are used to improve efficiencies. **Example Practices:** • Cluster asset management partnerships (implementing identical asset management at multiple neighboring utilities and sharing staff to maintain program support). • Multi-sector asset management relationships (e.g., with transportation sector). • Options purchasing for future, diversified source water supply (e.g., taking an option on the future purchase of a natural water storage source, like a quarry). • Fully developed enterprise asset management system.

Operational Resiliency

Scope: This core management area includes risk assessments, safety and security measures, all hazards disaster planning, emergency response and recovery, cyber security, and continuity of operations planning.

Level 1: Providing Adequate, Fundamental Services	**Practice Statement – Risk Assessment and Reduction Plan**: Risks to high-consequence assets are identified and reduced. **Example Practices:** • Risk assessment for high-consequence assets (i.e., those that would result in high public health or economic impacts if damaged). • Risk reduction plan containing countermeasures with prioritized list of mitigation projects (i.e., near- or long-term capital improvement projects). • Low-cost or near-term process improvement projects (e.g., fences and barriers around key utility facilities and infrastructure; doors and gates routinely locked; chemicals stored safely and securely, and properly disposed of; computers and network systems protected with passwords, and passwords changed routinely; abnormal conditions or activities reported by personnel; employee training in basic workplace safety practices and to actively monitor for abnormal or threatening situations and activities). • Cyber security measures (e.g., virus protection and firewall programs on all computers; electronic files and network systems regularly backed up). • Flood resilience measures (e.g., flood threats understood and practical mitigation options identified to protect critical assets). **Practice Statement – Emergency Response Planning**: Emergency Response Plan is developed containing basic policies and procedures. **Example Practices:** • Basic system information documentation (e.g., system maps and drawings) stored in secure on-site and off-site locations. • Emergency roles and responsibilities identification for utility personnel and local response partner agencies (e.g., law enforcement, fire, laboratories, public health agencies, and emergency management agencies). • General communication procedures (e.g., who activates the plan, order of notification, and contact information). • Training and exercise plan (to identify strategic goals and priorities for training and exercises). • Key utility response personnel training (in Incident Command System (ICS) and a plan to implement ICS during an emergency). • Critical customer needs and requirements identification and associated response protocols. **Practice Statement – Recovery and Mitigation**: General awareness of mitigation and recovery activities, projects, and funding is in place for efficient system and services restoration. **Example Practices:** • Local and state officials identified that would be involved in recovery (e.g., local community planners and State Hazard Mitigation Officers).

	• Local and state official coordination (e.g., local community planners and State Hazard Mitigation Officers). • Knowledge acquisition to understand options for resilient projects, concepts, and strategies, such as flood-proofing and relocating at-risk assets. • Awareness of the required documentation and application processes for federal funding programs.
Level 2: **Optimizing** **Operations and** **Services**	**Practice Statement – Risk Assessment and Reduction Plan:** Increase capacity to understand and detect threats to the system, risks to all major assets are identified and reduced, and all hazards risk management needs are fully integrated into broader utility planning and investment activities. **Example Practices:** • Risk assessment for all major assets (e.g., physical and cyber security, and business activities), including assessments of consequences and failure potential. • Risk reduction plan with a prioritized list of risk mitigation projects that, if fully implemented, would achieve acceptable risk levels for all major assets (e.g., hardening for facilities vulnerable to security threats and natural disasters; electronic files and network systems regularly backed up; chemical delivery control; intruder detection systems). • Risk reduction plan integration with long-range and capital investment planning for other projects. • Understanding regional environmental risks (e.g., fires, floods, earthquakes, tornados) and their relationship to utility operations and infrastructure (updated and maintained as current). • Identification and analysis of a wide range of contaminants and their properties (e.g., through the Water Contamination Information Tool). • Continuous on-line instrumentation for establishing trends and detecting abnormal occurrences (e.g., for pH and chlorine) in the water distribution system. **Practice Statement – Emergency Response Planning**: The Emergency Response Plan is enhanced with additional capabilities and supported through more structured relationships with potential response partners. **Example Practices:** • Alternate water source identification and alternate water supply distribution plans. • Mutual aid agreements (e.g., partnerships with neighboring systems for emergency response planning, participation in Water and Wastewater Agency Response Network (WARN), membership in an integrated nationwide network of laboratories such as the Water Laboratory Alliance). • Risk communication procedures for issuing messages during an emergency. • Business continuity plan (for maintaining solid operations–financially, managerially, and functionally–after any incident). • Routine joint training with neighboring utilities and response partners (e.g., full-scale exercises, mutual aid response/requests). • Utility representation in local Emergency Operations Center. • Response resources organized according to the AWWA resource typing manual. **Practice Statement – Recovery and Mitigation:** Implementation of mitigation and recovery activities, projects, and funding is in place. **Example Practices:** • Recovery plan (developed through collaborations with local and state officials that would be involved in recovery, including establishing clear roles and responsibilities for key partners such as local community planners and State Hazard Mitigation Officers). • Retainer contracts with consultants and backup equipment acquisition. • Business preparedness and continuity plan (developed, tested, and maintained to continue basic business operations during and immediately after disruptive events).

	• SOPs for documenting pre- and post-disaster condition of key assets applying for the federal funding program. • Key resilient projects, concepts, and strategies implementation, such as flood-proofing and relocating assets at risk from extreme weather events.

Level 3: Diversifying Operations and Services for the Future	**Practice Statement – Risk Assessment and Reduction Plan:** Emergent risks to all major assets are consistently addressed. Proactive and specialized shifts in operational procedures and updated capital investment criteria are changed when necessary.
	Example Practices:
	• Monitor/scan proactively for modern and emergent threats, and real-time monitoring for threat progression (e.g., watershed monitoring networks that support progressive storm alert systems).
	• Integrated Water Quality Surveillance and Response System addressing potential contamination within the distribution system.
	• Regular research on emerging trends that could pose new threats to the system, including changing weather patterns (i.e., climate change risk assessment integrated into existing risk assessment and reduction plan) and contamination threats.
	• Diversification and redundancy for critical supply, distribution, and treatment functions (e.g., emergency interconnects or bulk loading stations).
	
	Practice Statement – Emergency Response Planning: Emergency Response Plan is enhanced with incident-specific Emergency Action Procedures (EAPs) for responding to a specific type of incident, and enhanced capability to test, exercise, and to refine the Emergency Response Plan is in place. Ability to respond to a full suite of unexpected events by implementing a comprehensive Emergency Response Plan.
	Example Practices:
	• Specific EAP's for incidents, including the following:
	  o Severe weather response (e.g., snow, ice, temperature, lightning, flooding, hurricane, tornado)
	  o Fire response
	  o Electrical power outage response
	  o Water supply interruption response
	  o Earthquake response
	  o Disgruntled employee response (e.g., workplace violence)
	• Reviewed and updated utility response plans based on training and exercise activities (e.g., operations-based drills, functional and full-scale exercises), operational changes, and lessons learned from emergencies.
	• Capability to respond to mutual aid requests in self-sufficient manner, including cross-training staff to support neighboring utilities in the event of a mutual aid request.
	• Integrated consequence management plans as part of a Water Quality Surveillance and Response System for responding to contamination within the distribution system.
	• Interstate mutual aid request response plan (through Emergency Management Assistance Compact).
	
	Practice Statement – Recovery and Mitigation: Ability to recover from a full suite of incidents through implementation of comprehensive mitigation and recovery activities, projects, and funding is in place.
	Example Practices:
	• Prepared to conduct long-term public health and environmental health monitoring after a contamination incident.
	• Advanced contracts and agreements to support continuity plan implementation when needed.

- Detailed decontamination decision-making framework (established for remediation/cleanup).
- Remediation techniques and remedial process for treatment works and contamination distribution/collection systems implementation ability.
- Climate adaptation plan prepared with internal utility and community partners (climate adaptation measures, such as increasing water supply storage capacity for droughts, establishing alternative power supply, and monitoring flood and event drivers).

Water Resource Adequacy

Scope: *This core management area covers water resources, including water productivity and water reliability.*

Level 1: Providing Adequate, Fundamental Services	**Practice Statement – Water Reliability:** Essential elements of future water demand (e.g., population growth, industry production) are understood and factored into utility strategic and capital planning. **Example Practices:** • Demand forecasting (e.g., population change and existing per capita water utilization rates). • Pollutant restriction ordinances (to prevent unsuitable pollutants entering water resources through ordinance enforcement). • Level of service targets (based on historical use of water - equivalent residential connection - wastewater, and re-use - equivalent irrigation connection) for use in planning for future adequacy. • Service area definition. • Source water assessment and protection program. (Identify potential sources of contamination). **Practice Statement – Supply and Demand Management:** Existing water sources treatment and distribution is optimized. **Example Practices:** • Real water loss tracking and management. • Peak hour demand management (e.g., treat water during off-peak hours). • Water conservation plan.
Level 2: Optimizing Operations and Services	**Practice Statement – Water Reliability:** Utility prepared to meet the water or sanitation needs of its customers for the reasonable future. **Example Practices:** • Single scenario supply and demand forecasting and analysis. • Demand management plan (in place to influence short- and mid-term timing and efficiency of use). • Drought management plan that triggers actions for rationing or other demand reduction measures. • Ecological uses forecasts. • Water re-use plan implementation with regional water and wastewater utilities. **Practice Statement – Supply and Demand Management:** Utility has a conservation strategy covering all water users in its system and has initiated water re-use initiatives. Water use optimization is integrated into utility operational strategy, with fundamental water conservation and re-use methods implemented. **Example Practices:** • Low-flow toilets and faucets incentives for customer water conservation. • Integrated water conservation and re-use master plan (applied to water and wastewater infrastructure, as well as long-term planning). • Incentives for low water demand landscaping.

	• Water conservation and re-use tactics for all facilities and infrastructure, and encouraged for customers (e.g., provide discounted rain barrels to customers). • Plan implemented, including specific targets, for water recycling/re-use. • Water re-use for landscaping at utility facilities and at other municipal properties.
Level 3: Diversifying Operations and Services for the Future	**Practice Statement – Water Reliability**: Utility has an integrated, long-term water resources management approach that has addressed the potential for uncertainty in supply and demand conditions and effectively balances commercial, residential, and ecological needs. Utility is an advocate for and supporter of regional, integrated water management (e.g., a "One Water" approach), stewardship initiatives, and has an integrated water and energy long-term management approach. **Example Practices:** • Long-term water supply and demand analysis that considers long- term historical supply trends (e.g., 100+ years) and uses multiple demand and supply scenarios to identify robust implementation options. • Watershed-based plan to address all water resource demands (commercial, industrial, residential, and ecological). • Watershed council that integrates urban, agricultural, industrial users for optimized water allocation. • Leadership and advocacy for a sustainability master plan with coordinated objectives for water-energy actions/efficiencies (e.g., use high-energy water treatment to mitigate water scarcity risk, such as desalinization, or minimize the use of energy to conserve natural resources and reduce greenhouse gas emissions). **Practice Statement – Supply and Demand Management:** Utility leads or participates in studies and planning for developing and estimating conservation potential of utilities over a defined planning period (e.g., 20 years). **Example Practices:** • Local or regional utility and regulator partnerships to estimate indoor and outdoor conservation potentials by customer type. • Utility account-level information aggregation to develop prioritized water conservation initiatives/plans based on potential water savings and costs associated with conservation. • Emerging treatment technology utilization for wastewater treatment and low-energy solutions for water reclamation and energy-focused resource production. • Watershed-based permitting strategy participation or advocacy to enable water quality trading and market credits (e.g., advanced wetlands mitigation credits), and water rights trading. • Nontraditional partnerships with rivers, oceans, or agricultural organizations to identify re-use opportunities.

Community Sustainability

Scope: *This core management area covers social, economic, and environmental impacts relevant to utility operations.*

Level 1: Providing Adequate, Fundamental Services	**Practice Statement – Social Stewardship:** Utility operations, particularly siting and construction, are managed to minimize social impacts on the community. **Example Practices:** • Preconstruction notifications to households. • Construction hours of operation policy considerate of household needs. • Right-of-way procedures to provide for household access. • Utility staff teams support to community fundraising events. • Truck traffic management to reduce community impacts. • Riverfront access provision. **Practice Statement – Economic Stewardship:** Utility operations, particularly siting and construction, are managed to minimize economic impacts on the community. **Example Practices:** • Preconstruction notifications to local business. • Customer access to businesses in construction areas carefully accommodated. • Host community benefit program. (Provide for lower utility rates for communities that host treatment infrastructure). **Practice Statement – Environmental Stewardship:** Utility is focused on ensuring compliance across all regulatory areas to ensure a solid foundation for limiting environmental impacts of operations. **Example Practices:** • Environmental impacts review of regular utility operations and construction projects. • Annual state of watershed data review (to maintain awareness of ecosystem trends). • Sewer system maintenance to reduce flooding and backup potential.
Level 2: Optimizing Operations and Services	**Practice Statement – Social Stewardship:** Utility conducts operations with a view for improving community social conditions. **Example Practices:** • Women- and minority-owned business contracting policies. • Social impact criteria inclusion in project selection methods (e.g., degree of minority community impact). • Citizens' new facility siting committee. • Workforce diversification policies. **Practice Statement – Economic Stewardship:** Utility conducts operations with a view for enhancing local economic opportunity. **Example Practices:** • Utility finance officer's engagement in utility efforts to think and act sustainably.

	• Local materials and services sourcing policy, with targets for volume of services and products sourced from local firms. • Local and regional community and economic development planning participation. • Supplemental environmental projects (undertaken in lieu of paying noncompliance fines). **Practice Statement – Environmental Stewardship:** Utility operations and investments use techniques that enhance environmental and ecological parameters local to its facilities and operations. **Example Practices:** • Tree planting along utility right-of-way areas. • Discharge management to aid local fish populations. • Well-defined sustainability requirements (established and tracked for all key products and services). • Pollution prevention plan. • Low or no carbon fuels for vehicle fleet. • Water re-use for environmental protection purposes (e.g., to cultivate native or endangered plant species). • Annual sustainability report to show commitment to and performance on utility-related sustainability targets (e.g., greenhouse gas reduction targets and renewable energy utilization targets). • Recycled materials and product purchasing policies (adopted to encourage buying products manufactured from recycled materials and using recyclable products when such products are available). • Cradle-to-cradle studies to support choosing capital improvement project materials.
Level 3: Diversifying Operations and Services for the Future	**Practice Statement – Social Stewardship:** Utility is an active participant and takes a leadership role in driving overall community social development activities. **Example Practices:** • Elementary, secondary, and post-secondary school partnerships for collaborative efforts in promoting curriculum in water industry careers. • Volunteer/paid intern programs, leading to career choices and workforce sustainability of critical utility jobs. • Utility strategic plan with Triple Bottom Line decision making to support and incorporate community sustainability interests and priorities. **Practice Statement – Economic Stewardship:** Utility is an active participant and takes a leadership role in driving overall community economic development activities and performance. **Example Practices:** • Collaborative agreements with local firms to identify and provide stewardship services (e.g., reused water for certain industries). • Reliable, resilient, affordable, and sustainable water services marketing to prospective industry. • Community leadership for promoting green job growth and workforce sustainability. • Biogas or electricity production for directed marketing to energy grids. **Practice Statement – Environmental Stewardship:** A utility-wide environmental stewardship plan is in place that integrates environmental sustainability programs with sustainability priorities of the broader community. Utility conducts operations and makes investments to support broader community sustainability and stewardship goals.

Example Practices:

- Natural treatment systems to create "ecological bridges" to water bodies.
- Watershed-wide forums on source protection and enhancement.
- Native plant restoration program (e.g., seed and grow native plants in utility's watershed).
- Leadership on sustainability planning for energy and subsequent greenhouse gas reduction initiatives.
- Greenhouse gas offset investments (e.g., tree planting on utility rights-of-way).
- Watershed ecosystem services protection through land conservation acquisitions.

Performance Measurement and Continual Improvement

Scope: *This core management area covers considerations taken by utilities when managing achievement and measuring continuous improvement of performance.*

Level 1: **Providing** **Adequate,** **Fundamental** **Services**	**Practice Statement – Performance Measurement:** Critical performance metrics are established and tracked/monitored to help ensure compliance and achievement of improvement objectives. **Example Practices:** • Standard, basic reports to utility management and regulatory agencies addressing compliance requirements (e.g., source water quality, drinking water contaminants, wastewater discharge, and residual compliance metrics reports). • Procedures or assigned monitoring and measuring activities to include correcting/reporting of any nonconformance (e.g., use AWWA Operational Guides to G-Series Standards to establish procedures and performance measures, where applicable). • Proper calibration and maintenance for equipment used to measure performance metrics. **Practice Statement – Change Management and Continual Improvement:** A management process is in place to review critical metrics regularly (e.g., monthly) and make assignments for improvement actions when needed. Periodic and comprehensive self-assessments are conducted to identify areas needing improvement. **Example Practices:** • Quarterly review to compare progress towards compliance targets and objectives and identify process or procedural changes with a focus on improvement. • Annual review of established compliance performance levels from existing programs. • Active management consideration of new programs/improvements needed to achieve and maintain compliance.
Level 2: **Optimizing** **Operations and** **Services**	**Practice Statement – Performance Measurement:** Performance metrics are established in support of level of service commitments made to regulators, rate payers, and the community and in support of continual improvement objectives of the utility. **Example Practices:** • Targeted improvement for specific level of service processes, practices, or procedures with determination of what information will be collected to objectively demonstrate improvement. • Standardized data for comparison (e.g., compare treatment costs between plants based on 1,000 gallons of water treated). • Level of service metrics benchmarking to industry standards where benchmarks are applicable and available. • Monthly (or more frequent) critical performance measurements review by process owners to objectively track and trend continual improvement. • Periodic management review of measureable objectives, targets, and program steps to authenticate measured improvement.

	• Explicit performance objectives, targets, and programs to reduce significant environmental and business risks to the utility. **Practice Statement – Change Management and Continual Improvement:** Annual cycle of continual improvement is explicitly established and documented with new/revised annual objectives, supported by regularly monitored/tracked metrics, with an annual review leading to any needed program and operational improvements. Explicit continual improvement management system(s) is implemented, with an innovative workforce trained in continuous improvement tools providing a culture of continual improvement and innovation. **Example Practices:** • Operational and business practice process improvement management framework implementation and maturation. Examples of frameworks include: o ISO 14001:2004 certified or compliant Environmental Management System o AWWA Operational Guide to AWWA Standard G400, Utility Management System o Baldrige Performance Excellence Program o Balanced Scorecard • Document controls to capture timely changes, track changes, and establish periodic review for documents. (Capture change with program documents, SOP's, work instructions, forms, and record revisions; also, identify distribution, controlled copy location, retention, and disposition of documents). • Voluntary improvement program participation (e.g., Partnership for Safe Water, American Public Works Association [APWA] Accreditation Program). • Process unit and process variable monitoring to identify improvement opportunities. • Semi-annual review and comparison of measured performance to established industry benchmarks, and establishing improvement action plans.
Level 3: Diversifying Operations and Services for the Future	**Practice Statement – Performance Measurement:** Integrated, automated system of metrics measurement and tracking is in place supporting substantial real-time tracking of key performance indicators. Metrics established and reported to support Triple Bottom Line performance across a range of environmental, economic, and social parameters. **Example Practices:** • Mobile applications to record and upload field data. • Performance measurements integration with control charts to monitor, control, and improve process performance over time by reducing variation and its source, and also achieving incremental sustainable improvements. • Externally oriented metrics for utility-led community performance measurement (e.g., watershed health indicators; municipal government or community energy efficiency and conservation strategy; community or government greenhouse gas inventory and reduction goals; and support of community sustainability master plan with specific utility activities and performance metrics). • Performance and sustainability metrics reports to utility management and stakeholders, including regulators and financial institutions when applicable. • Greenhouse gas emissions reports in accordance with standard industry-accepted protocols. • Technology-enabled improvement (e.g., power management at treatment plants and collection and distribution systems through the automated interpretation of electronically gathered data for control chart or process capability real-time monitoring). **Practice Statement – Change Management and Continual Improvement:** Existing continual improvement systems and culture is built upon by using continual improvement and complementary management systems across all aspects of operations, and an emphasis is placed on cultivating community thought leaders to develop and harness the intellect of human assets.

Example Practices:

- Lean and Six Sigma optimization methodologies (with supporting work teams) to deliver customer requirements without waste.
- Continuous improvement management framework (EMS, UMS, BPEP, etc.) integration with other relevant and complementary continuous improvement management systems (e.g., ANSI Z-12 Continuous Improvement Safety Management System, the ISO 50001 Energy Management Standard, and the ISO 9001:2008 Quality Management System).
- Continuous improvement tools training (e.g., for Lean, Six Sigma, and statistical tools).
- Capturing innovative ideas with incentivized formal suggestion and process improvement programs.
- Municipal sustainability planning leadership (municipal activities addressing Triple Bottom Line decisions, measurement of sustainability improvements, and continuous improvement/sustainability reporting).

Conclusion

The challenges facing water sector utilities are significant and changing rapidly. These challenges are also exacerbated by uncertain economic conditions and other competing local priorities. However, along with these challenges come a myriad of important opportunities

This document provides an important foundation for utilities across the water sector to address these challenges by improving their current operations and moving toward the goal of sustainable operations over time.

More importantly, EPA recognizes that ensuring the sustainability of our nation's water sector utilities and, by extension, our nation's water resources, can only occur if regulators, utilities, states, and other partners work collaboratively and with common purpose. The challenges facing us are too daunting to do otherwise.

Through this document and other efforts, EPA is committed to this collaboration going forward.

Appendix: Resources Directory

The following appendix is a resources directory designed to support water and wastewater utilities in implementing practices in the key management areas. The directory is not meant to be an exhaustive compilation, but rather is a starting place for users of this document. A few notes for users of the resource directory:

- Resources are organized alphabetically by title, with check marks to indicate which management area(s) each resource applies to.
- Resources that are designed specifically for use by small systems are also marked.
- There is a web link for each resource. *(Links to resources may change as they are updated)*
- All of the resources listed are free.

Description and Link	Utility Business Planning	Product Quality & Operational Optimization	Customer Satisfaction and Stakeholder Understanding & Support	Employee & Leadership Development	Financial Viability	Infrastructure Stability	Operational Resiliency	Community Sustainability	Performance Measurement and Continual Improvement	Small Systems
A Drop of Knowledge: The Non-Operator's Guide to Drinking Water Systems Explains in simple language the technical aspects of drinking water utilities from source to tap. It would be well used as an orientation and background for new small utility board members and community decision makers. http://www.rcap.org/sites/default/files/rcap-files/publications/RCAP-Non-operator%27s%20Guide%20to%20DRINKING%20WATER%20Systems.pdf			X							X
A Drop of Knowledge: The Non-Operator's Guide to Waster Systems Explains in simple language the technical aspects of wastewater utilities from source to tap. It would be well used as an orientation and background for new small utility board members and community decision makers. http://www.rcap.org/sites/default/files/rcap-files/publications/RCAP-Non-operator's%20Guide%20to%20WASTEWATER%20Systems.pdf			X							X
Asset Management: A Best Practices Guide Designed to help owners, managers, and operators for small water systems to understand: what asset management means; the benefits of asset management; best practices in asset management; and how to implement an asset management program. http://epa.gov/ogwdw/smallsystems/pdfs/guide_smallsystems_assetmanagement_bestpractices.pdf						X				

Description and Link	Utility Business Planning	Product Quality & Operational Optimization	Customer Satisfaction and Stakeholder Understanding & Support	Employee & Leadership Development	Financial Viability	Infrastructure Stability	Operational Resiliency	Community Sustainability	Performance Measurement and Continual Improvement	Small Systems
Asset Management: A Handbook for Small Water Systems Designed for owners and operators of small community water systems (public or private). It presents basic concepts of asset management and provides the tools to develop an asset management plan. http://epa.gov/safewater/smallsystems/pdfs/guide_smallsystems_asset_mgmnt.pdf						X				X
Check up Program for Small Systems (CUPSS) CUPSS is a free, easy-to-use, asset management tool for small drinking and wastewater utilities. Use CUPSS to help you develop: - A record of your assets - A schedule of required tasks - An understanding of your financial situation - A tailored asset management plan http://water.epa.gov/infrastructure/drinkingwater/pws/cupss/	X				X	X				X
Climate Ready Water Utilities Toolbox Provides access to resources containing climate-related information relevant to the water sector. The Toolbox contains highlighted resources, which are organized into categories to help guide the user to the most relevant information. CRWU resources are updated frequently. http://www.epa.gov/safewater/watersecurity/climate/toolbox.html	X				X	X	X	X		
Communicating the Value of Water A guidebook to help drinking water utilities effectively communicate the value of their tap water. Identifies the benefits of communicating effectively and consistently to constituents about the value of water. Includes key messages, marketing tools, and case studies. http://www.waterrf.org/PublicReportLibrary/91222.pdf										
Confronting Climate Change: An Early Analysis of Water and Wastewater Adaptation Costs Details the impacts that climate change can have on wastewater and drinking water utilities, as well as the adoption costs for these critical facilities. http://www.nacwa.org/images/stories/public/2009-10-28ccreport.pdf	X				X	X	X	X		

Description and Link	Utility Business Planning	Product Quality & Operational Optimization	Customer Satisfaction and Stakeholder Understanding & Support	Employee & Leadership Development	Financial Viability	Infrastructure Stability	Operational Resiliency	Community Sustainability	Performance Measurement and Continual Improvement	Small Systems
Cybersecurity Guidance and Tool Resources designed to provide actionable information for utility managers and operators based on their use of process control systems. These resources complement the national-level actions that have resulted from Executive Order 13636 - Improving Critical Infrastructure Cybersecurity. http://www.awwa.org/resources-tools/water-and-wastewater-utility-management/cybersecurity-guidance.aspx							X			
Effective Utility Management: A Primer for Water and Wastewater Utilities The *Primer* presents a framework for water and wastewater utility managers to use when assessing the effectiveness of their utility based on a series of 10 Attributes of Effectively Managed Utilities and Keys to Management Success. http://www.watereum.org/WorkArea/DownloadAsset.aspx?id=100	X								X	
Emergency Response Plan Guidance for Small and Medium Community Water Systems Provides guidance on developing or revising emergency response plans for small- and medium-sized community drinking water systems, to comply with the Public Health Security and Bioterrorism Preparedness and Response Act of 2002. http://water.epa.gov/infrastructure/watersecurity/upload/2004_04_27_watersecurity_pubs_small_medium_ERP_guidance040704.pdf							X			X
Emergency/Incident Planning, Response, and Recovery Includes guidance documents and other resources on resilience topics, including: training and exercise, mutual aid and assistance, coordination with state agencies, all-hazard planning, emergency response plan development, risk and crisis communication, and more. http://water.epa.gov/infrastructure/watersecurity/emerplan/index.cfm#te2							X			
Energy Efficiency Best Practices for North American Drinking Water Utilities Includes a compendium of best practices for energy efficient design and operation of water industry assets. Compendium includes successful strategies to help water utilities reduce energy consumption in water transmission, treatment, storage, and distribution. http://www.waterrf.org/Pages/Projects.aspx?PID=4223		X				X				

Description and Link	Utility Business Planning	Product Quality & Operational Optimization	Customer Satisfaction and Stakeholder Understanding & Support	Employee & Leadership Development	Financial Viability	Infrastructure Stability	Operational Resiliency	Community Sustainability	Performance Measurement and Continual Improvement	Small Systems
Energy Efficiency for Water Utilities Provides links to several guides and tools for tracking and understanding water utility energy use. http://water.epa.gov/infrastructure/sustain/energyefficiency.cfm		X								
Energy Star for Wastewater Plants and Drinking Water Systems A tool for plant managers to assess and track energy use, energy costs, and associated carbon emissions. Allows benchmarking against other systems. http://www.energystar.gov/index.cfm?c=water.wastewater_drinking_water		X								
Ensuring a Sustainable Future: An Energy Management Guidebook for Wastewater and Water Utilities Provides water and wastewater utility managers with step-by-step methods to identify, implement, measure, and improve efficiency and renewable opportunities at their utilities. http://www.epa.gov/owm/waterinfrastructure/pdfs/guidebook_si_energyman agement.pdf		X				X		X		
Establishing Public-Private Partnerships for Water and Wastewater Systems Describes the conditions for when to form a public-private partnerships and how to form and manage them to meet water and wastewater needs. http://www.nawc.org/uploads/documents-and-publications/documents/document_567764ad-b69f-4715-bc5d-eaa32c304fdd.pdf			X					X		
Federal Funding for Utilities - Water/Wastewater - in National Disasters (Fed FUNDS) Fed FUNDS provides tailored information to water and wastewater utilities about applicable federal disaster funding programs. The Fed FUNDS web pages address national-level disasters, but can also apply to large-scale and local disasters that result in service interruptions and significant damage to critical water/wastewater infrastructure. http://water.epa.gov/infrastructure/watersecurity/funding/fedfunds/index.cfm					X	X	X			
Financing Alternatives Comparison Tool (FACT) A financial analysis tool that calculates and compares the costs of various financing options for water quality projects. http://www.epa.gov/owm/cwfinance/cwsrf/fact.htm	X				X					

Description and Link	Utility Business Planning	Product Quality & Operational Optimization	Customer Satisfaction and Stakeholder Understanding & Support	Employee & Leadership Development	Financial Viability	Infrastructure Stability	Operational Resiliency	Community Sustainability	Performance Measurement and Continual Improvement	Small Systems
Formulate Great Rates: The Guide to Conducting a Rate Study for a Water System A guide to developing a fair and equitable rate structure in a small drinking or wastewater system. http://www.rcap.org/sites/default/files/resource_attachments/rcap_basics_of_financial_management_0.pdf	X				X					X
Getting in Step: Engaging and Involving Stakeholders in your Watershed Provides tools needed to effectively engage stakeholders to restore and maintain healthy environmental conditions through community support and cooperative action. Can help utilities to involve stakeholders in local or regional watershed efforts. http://cfpub.epa.gov/npstbx/files/stakeholderguide.pdf			X					X		
Green Infrastructure Provides background information and resources on green infrastructure strategies. http://water.epa.gov/infrastructure/greeninfrastructure/index.cfm	X					X		X		
How to Develop a Multi-Year Training and Exercise Plan Provides background on different types of training and exercise, describes the importance of a training and exercise plan, provides a multi-year plan template, and includes attachments with example plan documents, exercise resources, and planning resources. http://water.epa.gov/infrastructure/watersecurity/emerplan/upload/epa816k11003.pdf							X			
Large Water System Emergency Response Plan Outline Provides guidance to assist community water systems in developing or revising emergency response plans to comply with the Public Health Security and Bioterrorism Preparedness and Response Act of 2002. http://water.epa.gov/infrastructure/watersecurity/upload/2004_04_01_watersecurity_erp-long-outline.pdf							X			
Managing Money: State SRF Short-Term Investing Focuses on strategy development for SRF managers. It looks at shorter-term investment goals, where consideration of liquidity and accessibility are controlling factors. http://www.cifanet.org/newsPDF/m12.pdf	X				X					

Description and Link	Utility Business Planning	Product Quality & Operational Optimization	Customer Satisfaction and Stakeholder Understanding & Support	Employee & Leadership Development	Financial Viability	Infrastructure Stability	Operational Resiliency	Community Sustainability	Performance Measurement and Continual Improvement	Small Systems
NIST Cybersecurity Framework The *Framework for Improving Critical Infrastructure Cybersecurity* includes standards, guidelines, and practices to promote the protection of critical infrastructure and manage cybersecurity-related risks. http://www.nist.gov/cyberframework/							X			
Optimizing the Water Utility Customer Contact Center Identifies best practices, processes, and technologies for water utility customer contact center operations to optimize the contact center as a utility-wide resource for communications. Identifies key components and characteristics of the customer contact center of the future. http://www.waterrf.org/PublicReportLibrary/4100.pdf			X							
Performance Benchmarking for Effectively Managed Water Utilities A tool and corresponding resources to help water and wastewater utilities evaluate their current and desired levels of performance related to the Effective Utility Management 10 Attributes. This tool and resources are aligned with the *EUM Primer* and provides a structured process to help utilities conduct a self assessment on any or all of the 10 Attributes. http://www.waterrf.org/Pages/Projects.aspx?PID=4313	X	X	X	X	X	X	X	X	X	
Planning for Sustainability: A Handbook for Water and Wastewater Utilities Describes steps that utilities can take to enhance their existing planning processes to ensure that water infrastructure investments are cost-effective over their lifecycle, resource efficient, and support other relevant community goals. http://water.epa.gov/infrastructure/sustain/upload/EPA-s-Planning-for-Sustainability-Handbook.pdf	X				X				X	
Rural and Small Systems Guidebook to Sustainable Utility Management The *Guidebook* uses the Effective Utility Management (EUM) framework, and is tailored to the needs of rural and small systems. It is designed to help them become more successful and resilient service providers, and includes a utility self assessment. http://water.epa.gov/infrastructure/sustain/upload/SUSTAINABLE-MANAGEMENT-OF-RURAL-AND-SMALL-SYSTEMS-GUIDE-FINAL-10-24-13.pdf	X	X	X	X	X	X	X	X		X

Description and Link	Utility Business Planning	Product Quality & Operational Optimization	Customer Satisfaction and Stakeholder Understanding & Support	Employee & Leadership Development	Financial Viability	Infrastructure Stability	Operational Resiliency	Community Sustainability	Performance Measurement and Continual Improvement	Small Systems
Strategic Planning: A Handbook for Small Water Systems A strategic planning handbook and workbook for small water systems. http://www.epa.gov/ogwdw/smallsystems/pdfs/guide_smallsystems_stratplan.pdf	X				X					X
Survival Guide: Public Communications for Water Professionals Helps utilities learn how to effectively communicate with their communities and customers. Provides an overview focused on learning the basics of public communication. www.wef.org/WorkArea/DownloadAsset.aspx?id=7120			X							
The Basics of Financial Management for Small Community Utilities A primer and how-to guide that is ideal for a board member of a drinking or wastewater utility who needs to understand the financial aspects of a small utility's operations. http://www.rcap.org/sites/default/files/resource_attachments/rcap_basics_of_financial_management_0.pdf					X					X
The Effective Utility Management Resource Toolbox Provides a compilation of resources from collaborating associations and agencies on the EUM effort, and is organized according to the 10 Attributes and five keys to management success. Select an attribute or management key to learn more and see the available resources. http://www.watereum.org/resources/resource-toolbox/	X	X	X	X	X	X	X	X	X	
The Water Resources Utility of the Future: A Blueprint for Action Presents the clean water industry's vision for the future, as well as a series of actions that will help deliver this vision. The Utility of the Future will transform the way that traditional wastewater utilities view themselves and manage their operations, including their relationships with communities and their contributions to local economies. http://www.nacwa.org/images/stories/public/2013-01-31waterresourcesutilityofthefuture-final.pdf	X	X	X		X		X	X		
Utility Finance Knowledge Portal Includes resources on revenue, financial planning, board and customer communication, and cost control. http://www.waterrf.org/knowledge/utility-finance/Pages/default.aspx	X		X		X					

Description and Link	Utility Business Planning	Product Quality & Operational Optimization	Customer Satisfaction and Stakeholder Understanding & Support	Employee & Leadership Development	Financial Viability	Infrastructure Stability	Operational Resiliency	Community Sustainability	Performance Measurement and Continual Improvement	Small Systems
Value of Water Coalition Public education materials about the importance of clean, safe, and reliable water to and from every house and community, to help ensure quality water service for future generations. www.thevalueofwater.org			X					X		
Water & Wastewater Pricing - Introduction Website provides information on water and wastewater pricing; explains the concepts of pricing and water conservation; and supplies tools, guides, and reports on pricing. http://water.epa.gov/infrastructure/sustain/Water-and-Wastewater-Pricing-Introduction.cfm	X				X					
Weather & Hydrologic Forecasting for Water Utility Incident Preparedness and Response Includes resources for national weather hazards, national forecast charts, flood risks, drought monitors, and more. http://water.epa.gov/infrastructure/watersecurity/emerplan/upload/epa817f13005.pdf							X	X		
Work for Water Campaign promoting water careers as both professionally fulfilling and aligned to the greatest public health and environmental causes of our day. Includes resources for recruitment and retention, as well as management strategies. www.workforwater.org				X						
Workforce Planning for Water Utilities Frames the issues of recruiting, training, and retaining drinking water utility operators and engineers. Identifies short-term and long-term strategies that can be implemented by individual utilities and by the industry to address workforce planning issues. http://www.waterrf.org/PublicReportLibrary/91237.pdf				X						
Workforce Sustainability Provides resources for enhancing the image of water careers. Resources include information on workforce sustainability, upcoming events, relevant reports, and opportunities for involvement. http://www.wef.org/AWK/pages_cs.aspx?id=589				X